MARKETING ORGANIC GRAIN

A Farmer's Guide

MARKETING ORGANIC GRAIN

A Farmer's Guide

by John A. Bobbe

For my wife Terry

ACKNOWLEDGMENTS

I owe special thanks to Carmen Fernholz and Don Rudolph for reading the first manuscript draft. They provided valuable suggestions, ideas and corrections. As I think of the many organic farmers who have helped me learn the ins and outs of marketing organic grains, Oren Holle, Tom Bilek, Carmen Fernholz and the late Kevin Brussell come to mind. This is hardly a complete list—that would take most of my book!

I also want to thank the folks at Levins Publishing for helping me put this book together. Dick Levins suggested that I write this book and encouraged me at every step of the way. Jane Dickerson did the final edit and Alexandra Erickson supervised the design process.

Terry, my wife of 41 years, has always understood and supported my passion for farming, the land, and living in rural America. My future is my three children and their spouses, Elizabeth (Tom), Michael (Jennifer) and Stephen (Hannah) and my grandchildren, Sonya, Mitchell, Emilia, Eavan, Cael, Miera, Griffin and Harper.

A Lifelong Journey
to the Other Side of the River

My life's story is not unlike that of many other kids who grew up in rural America during the 1950's and 60's. Our family had a diversified dairy and livestock farm in central Wisconsin. We were the German family that moved into the Norwegian Scandinavia Township from the German-Polish community of Amherst just to the west.

My ancestors followed the great migration to America in the 1850's from Saxony, Germany. Eventually they settled in Amherst, Wisconsin, a small town just east and a bit south of Stevens Point. After trying their hand at homesteading in North Dakota, my great-grandparents returned to the Amherst community in 1902.

My grandparents started farming about a mile down the road from the original Bobbe homestead in Amherst. Times were tough leading up to and through the Great Depression; they lost everything when the Land Bank refused to refinance their loan. In 1942, my grandfather bought the 172-acre farm where my wife and I currently live. The years that followed were good ones for farming and the farm was paid for by the time my parents took over in 1950.

We always milked about 35 or 40 cows in a stanchion barn. We also had hogs, a small herd of Angus beef cattle, sheep, and chickens for both meat and eggs. Grandfather still kept his threshing business until combines took over in the early 1960's. He also had a sawmill where neighbors brought their logs for sawing into lumber. We still own and operate that sawmill today. I can remember going down the road a quarter-mile or so to the one-room Riverview School for the first three years of my education. Then we were bussed to Scandinavia to the four-room school and eventually eight miles to Iola for high school.

The floodgates of agricultural technology were starting to open wide. The herbicide Atrazine first made its appearance in the late 1950's. Atrazine, in combination with the slower-acting Simazine, did wonders on the heavy muck and peat ground we had sod-busted and ditched with government payments and blessings. I remember salesmen demonstrating how safe Atrazine could be by drinking a glass of it mixed with water at sales meetings. The herbicides 2, 4-D and 2,4,5-T, which we would later know as components of Agent Orange, were regularly used on some farm fields. The Green Bay and Western Railroad, which cut through the farm on the edge of the "back 40," also used those chemicals for brush control.

My parents survived the horrendous years of low prices and mass farm exodus during the years Ezra Taft

Benson served as President Eisenhower's secretary of agriculture. Farm prices stayed low into the 1960's and my father, finally fed up with working for nothing, started attending county farm meetings held by an up-start group called the National Farmers Organization. With considerable encouragement by a cousin, Dad joined NFO in 1963. So, I, barely a teenager at the time, started going to farm meetings and even writing letters for publications like *Hoard's Dairyman* about the plight of farmers like us.

After high school, I majored in agribusiness at the University of Wisconsin-River Falls and graduated in 1973 with a Bachelor of Science degree. My summer job was with the National Farmers Organization. Pete Nagel, a farmer from Blue Earth, Minnesota, hired me to help organize farmers in a number of counties in Wisconsin. I spent two summers going down the road talking to farmers about their concerns and the need to work together to solve the problem of low farm prices. During my third summer, I ran two livestock collection points, taking in cattle and hogs, weighing, sorting, and dispatching truckloads to meatpackers. My final summer after graduation, I worked as a dairy field man, contacting farmers about group action as well as doing milk quality work such as sampling and testing.

During the early part of my senior year at River Falls, my major professor was Dr. Gary Rohde. He suggested I consider graduate school. More college had never crossed my mind. After all, I was the first to graduate with a bachelor's degree in agriculture in all the generations of our family. My father did attend the University of Wisconsin-Madison Farm Short Course which was a 1-year program. With Dr. Rohde's guidance and help, I looked at the University of Wisconsin-Madison and several colleges a lot farther from home. In December, 1972, I made a trip to Columbia, Missouri, to talk with professors and students. Upon my return home, I found a letter in my mailbox from the department chairman offering me a graduate assistantship in the Department of Agricultural Economics. Missouri had a national reputation of advocating for family farms and favorable farm policies to foster them.

I spent 18 months at the University of Missouri studying both agricultural economics and rural sociology. My graduate committee suggested I focus my research on whether dairy programs like the one used by NFO could influence what was then called the Minnesota-Wisconsin Series Price, that is, the base price that farmers were paid for the milk they produced. My research showed that farmers acting together could indeed influence the price they were paid for milk. This got the attention of the United States Department of Agriculture. At the time, its official position was that there was nothing farmers could do about the Series Price, or, to use the agency's words, "farmers were price takers, not price makers." However, even after my conclusions were watered down under political pressure, USDA could not outright refute a lesson that makes as much sense today as it did then—farmers, strategically marketing together, can influence prices in ways that lead to better lives for them, their families, and their communities.

Right out of graduate school with a Master's Degree in Agricultural Economics, I was hired as an economist and assistant director of the NFO dairy program. In this position I performed economic analysis, worked on NFO's large anti-trust lawsuit in dairy and coordinated about 150 employees from Lake Michigan to the State of Washington in milk marketing for NFO's dairy producers.

Four years later, I took a break from the stressful work of organizing farmers and began a ten-year stint as an instructor at the Northeast Wisconsin Technical College in Green Bay. I taught what at the time was called Farm Training and was also involved with the on-campus agribusiness program. I had a commercial applicator license to teach farmers how to use every chemical imaginable and help them figure out their chemical fertilizer needs. Through all of this, I was having more and more trouble seeing how farmers were ever going to make enough to pay for all the new technology that chemical companies were throwing their way.

In 1986, I attended my first "sustainable agriculture" field day on the Carl Pulvermacher farm in Lone Rock, Wisconsin. After 10½ years of teaching farmers how to use what we were learning was "dangerous stuff," I was ready to learn more about new and better ways to farm. I spent my next four years organizing farmers in Door and Kewaunee Counties in Wisconsin into a sustainable farming network. From there, I spent about five years managing a grazing network that spanned the Great Lakes Basin from Minnesota to New York, including farms in Ontario. The project was to encourage more farmers to keep their land in grass and thereby reduce the flow of fertilizer and herbicide chemicals into the Great Lakes.

In 1996, I became National Farmers Organization assistant to the president. One of my assignments the following year was to look at how organic food-grade soybean farmers could organize in ways that would do something about the low prices they were getting. The more we worked with the organic soybean farmers, the more we realized that all organic farmers, not just those growing soybeans, faced the same problems of low prices and needed the same solution of group marketing. Things started coming together, and by 2001 I was serving part-time as the Executive Director of the newly formed "Organic Farmers' Agency for Relationship Marketing," or "OFARM."

A good friend of mine likens coming to organic agriculture from the conventional world to crossing the river. I made that crossing in July of 2001 and never once have I looked back. In the years since, I have been helping farmers make that crossing to the organic shore where expensive and dangerous chemicals are replaced by fairly-priced crops and livestock products.

In this little book, I want to help you make that crossing, too.

MARKETING ORGANIC GRAIN

A Farmer's Guide

CHAPTER ONE

Organic Is the Future of American Agriculture

In business and industry, entire research and marketing departments study the latest trends in what customers want and what trends will drive future purchases. While farmers don't have that luxury, they can learn a lot from what others are saying. Two major retailing entities are trading places in a horse race as to who sells the most organic and natural foods. Both exceed $13 billion annually in these sales categories and have hundreds of stores in the U.S. and Europe.

I was lucky enough to hear a global executive of one of those companies speak at a USDA conference in March of 2015. Here are some highlights of what he said:

- Consumers are demanding more information about a product's ingredient list, provenance, manufacturing process, shipping, storage, and safety testing.

- Accentuate what's *not* in a product as much as what is.

- The market for "free from" goods is growing rapidly.

- Prove your commitment to product safety by enforcing and publicizing strict internal standards.

He went on to cite data that dollar sales for non-GMO products at the retail level increased well over 400 percent from 2010 to 2014. Organic sales during that time increased in the high double-digits. Combined non-GMO and organic sales increased astronomically. Meanwhile, growth in traditional aggregate grocery store sales has been crawling along at less than 2.5% per year.

Stagnant sales are not the only reason conventional farmers are looking at organic agriculture. Prices for conventional agriculture are on such a roller coaster that planning for the future is all but impossible. In the past few years, conventional corn has sold for as high as $8.00 per bushel and for as low as $3.00. Conventional milk has sold for over $20.00 per cwt. and for less than $12.00. Financial stress is inevitable in an economic environment with such unstable prices.

The food company executive I heard speak concluded by saying that organic is the future of agriculture and that there is no other trend. I agree completely with that statement. I'm glad you are considering or are already part of American agriculture's new future— organic farming. I think you will be happy with a decision to go organic if that is what you ultimately decide to do. But before you jump, there are some things you need to think about.

One of the key differences between conventional and organic farming is that organic farming requires more day-to-day management. While one conventional farmer may use pretty much the same varieties, chemicals, and fertilizers as his or her neighbor one county over, the organic farmer will likely have developed a set of varieties, rotations, planting dates, and cultivation methods unique to his or her farm. Organic farming usually requires more labor. Finding people with the right skills at the right time can be a challenge. For these, and many other reasons, you will learn firsthand that patience is a major virtue in organic farming.

Organic farm products usually sell for substantially more than conventional products, but that benefit comes at a price that is partly measured in additional paperwork. Everything you send off your farm for sale to be sold as organic will require the necessary paperwork and documentation.

Another important difference between organic and conventional farming is that an organic certification farm plan is going to require a crop rotation. The organic rotation goes far beyond the corn-soybean system used by conventional farmers. An organic grain farmer's rotation will usually include a range of potential crops, including the major ones like corn, wheat, and soybeans. Many organic rotations also include oats and barley as well as other small grains and legumes as a source of nitrogen.

Before I go on, I want to emphasize that GMO varieties and most pesticides are not allowed in organic farming. Pollen and chemicals drifting onto your farm from neighboring fields can result in your crops being kept out of the organic market and sold into the conventional market at much lower prices. You will have to develop a plan to keep your organic crops "clean."

This a lot to take in all at once, so don't think you need to convert your whole farm at one time. One of the most common routes to taking a farming operation to 100% organic is to begin small with 40 acres. Learning to manage on that scale is less risky than the whole farm approach of going in all at once. I have seen a number of farmers go so-called "cold turkey" and be successful at it, but the learning curve was incredibly steep and often costly.

You should also be aware of the many organizations that can give you the on-going help you will need to be successful in organic farming. On the production side, some programs can pair you up with an experienced organic farmer who has already been through the mill of getting there. The Midwest Organic Sustainable and Organic Services (MOSES) offers one of the best programs I have seen. You can find more information at www.mosesorganic.org.

You also will need help marketing your crops. In your transition to organic farming, you may be able to sell some crops as non-GMO. Once your crops are certified organic, other special considerations arise that will be new to you. I hope you will consider joining one of the OFARM-affiliated cooperatives so you can get the marketing expertise you will need to make the most from the crops you grow.

I wish you every success on your road to becoming part of American agriculture's organic future, and hope we have other opportunities to talk about marketing strategies.

CHAPTER TWO

Organic Grain Farming from Planning to Paycheck

Conventional farming is prescription farming. You simply order and buy the seed, till, plant the crop, and read the directions as to how to mix and apply the herbicide or hire it done. Then it's watching and waiting until harvest time. On-farm and off-farm storage is readily available. And, more than likely, you can either forward contract or price your grain or forage at almost any time throughout the year.

On the other hand, organic farming requires much more intense management and decision-making. Choosing which crops to plant is a complex decision and you must know the approved list of allowable materials published by the Organic Materials Review Institute (www.omri.org). More often than not, doing things in a timely manner is critical. Learning how government programs apply to organic farming takes time. University experts to back you up when you go organic are often hard to find.

Here is a chronological checklist of steps to consider as you plan, produce, and market organic grains:

1. Investigate potential markets, including specifications, standards, grades, quality requirements, and prices so you can put together a financial plan before purchasing any seed. A good set of data is available from the University of Minnesota at http://www.finbin.umn.edu/. It includes specific data for organic crop enterprises from 2006 to present.

2. Developing and following an organic farm plan that goes from A to Z on everything you do or will do is part of the organic certification process. It needs to be updated as you go along through the year and modified as necessary. This involves extensive record keeping for each field, including necessary receipts for crop inputs such as seed and approved materials used in growing the crop.

3. An organic farm plan will include a crop rotation. Nitrogen is probably one of the most difficult nutrients to supply. Therefore, legume crops such as soybeans, alfalfa and clover are a key component of any organic crop rotation for growing corn. If at all possible, include some sort of livestock enterprise into your system as a source of nutrients from manure.

4. You must make every attempt to obtain certified organic seed, even if the price is high. If organic seed is not available, organic regula-

tions require you to make a good faith effort to locate the seed you need, and it cannot be treated. Start looking for seed early so you will have the best chance of getting the variety and type of seed your plan requires, not only seeds for major crops, but also those for green manure plow down crops.

5. Soil nutrient management and fertility require careful planning. Good soil organic matter is a key component. Green manure crops plowed down, approved fertilizers, compost, livestock manure, and chicken manure are good sources of nutrients. Other options for growing crops include foliar feeding of liquid fish fertilizer. Just be sure that everything you apply to any crop on your organic farm is listed on the approved organic materials list.

6. Wherever there is a chance of pollen or pesticide drift from conventional crops on neighboring fields, a general rule is that buffer strips at least 30 feet wide must be planted. Crops from buffer strips must be harvested and sold conventionally. It's best to consult your organic certifier about the specifics of when and where buffer strips are required.

7. Time of planting is often critical. The general concept for row crops is to let the soils warm up and use tillage to kill the first generation of

weeds. Warmer soil also allows for faster seed germination. This, along with a good supply of nutrients, gets the crop off to a fast start and makes weed control easier. In the case of corn, planting to time pollination to occur later than pollination in neighboring conventional corn fields will reduce potential GMO contamination.

8. Weed control is a major consideration in organic row crop production. Options here include rotary hoeing row crops once and even twice within a few days of planting. This is usually followed by mechanical cultivation and in some cases flame cultivation. Weather is often a critical factor in staying ahead of weeds, so keep your equipment available and well-maintained. In the case of forages and small grains, cutting and harvesting or using the crop as green manure may be an option.

9. Insect infestations can be a problem. Well balanced soils that promote healthy plants can help reduce insect impacts. In some crops, approved spraying of approved oils may also assist in pest control. The last resort is always cutting and plowing the crop down as green manure or as forage. However, the main objective is to have a harvestable crop as you originally intended.

10. Harvesting in a timely manner to maintain maximum quality of grains or forages harvested for hay is important. Any harvesting equipment must be free of GMO residue that may have been picked up during harvesting buffer strips or conventional crops. Equipment should be maintained and properly adjusted to maximize the amount of grain or forage harvested.

11. More often than not, commercial off-farm storage for organic crops is hard to find. Storing a crop on-farm requires bins to store the crop and keep it in proper condition with adequate air and drying capabilities. Strongly consider full floor aeration in any new construction and when retrofitting existing bins. Another highly recommended storage bin feature is to have reasonably rapid load out equipment when it comes to shipping the grain, especially if commercial trucking is required.

12. At some point during the growing season, there will be an on-farm inspection from your certifier. Your inspector will expect to see documentation for crops from each field and in each bin. You must also document crops from each year separately. Proper sampling techniques and necessary paperwork as recommended by your marketer or buyer must also be maintained.

Above all, patience is a key component in growing good organic grains and forages. It takes patience as you watch your conventional neighbors rush to beat Mother Nature to get their crops in early. For example, your best ally is working with Mother Nature by using warmer soils to achieve faster germination to your advantage. Learn to think of your weed control equipment as enduring investments and be grateful that you no longer have nothing to show for annual herbicide purchases except for big hits to your checkbook.

Finally, never stop looking for new ideas and practices. Two good sources of information I use all the time can be found at www.mosesorganic.org and www.northcentralsare.org in their publications and fact sheets sections. I'm sure you will find other good sources, too, as you get more familiar with the world of organic farming.

Well, not quite finally. This book is focused on organic grain marketing and getting paid a fair price for what you sell. However, a couple of other components should be considered in your overall marketing plan. Organic cost share programs and federal crop insurance programs are important, but they change so often that anything I write here might not be accurate by the time you read it.

There are a number of state and federal programs that can pay for at least part of certification and other farming practices that might fit into your overall strategic plan. Your local Farm Service Agency (FSA) office and your state department of agriculture can help you learn more.

The worst-case scenario for any marketing plan is to have a drought, hail, or some other disaster leave you with no crop to sell. Talking to other organic farmers about whether and how they insure their crops will be a good source of information for you. Chances are that in talking with them you can also find an insurance agent with a good understanding of federal crop insurance and the specific aspects of crop insurance that apply to organic crops.

CHAPTER THREE

Choosing the Right Crops
for Your Market

In conventional agriculture, you choose the crop you want to grow in the current season. Maybe this season you think corn, soybeans, wheat, or a specialty crop is the way to go. The crops you select can easily be determined by some enterprise budgets for the greatest profit potential. The important thing is that you are planning one year, sometimes two years, out at a time in conventional agriculture.

Choosing the right crops for your market is different in organic production. You start by determining your crop rotation. A crop rotation is required as part of your organic farm plan. You can alter the plan and rotation due to weather and soil conditions or other factors, but it is unwise to simply tear up the rotation to grow for a market and then be faced with how to meet the requirements of your rotation. Not meeting the organic requirements could result in decertification.

Two friends of mine who farm in two different states and have spent over 30 years each in organic farming in the Upper Midwest have found rotations that work for them. I only cite these as examples. One of them has a six-year rotation that includes four crops. It begins with two years of alfalfa, a year of soybeans, followed by one year of corn, another year of soybeans and, in the final year of the rotation, he plants oats under-seeded with alfalfa. Then the cycle repeats. My other friend also has a six-year rotation, but the crops in it are different. He begins with one year each of corn and soybeans, followed by a year of small grains (flax, wheat, barley, oats) under-seeded or followed by seeding clover (one or two years) or three years of alfalfa.

As I talked with one of these friends about his rotation, what struck me was how he decided which crops to plant. He said he keys in on the small grains in the first year of the rotation. He talks to his organic grain marketer about which of the small grains might have the most profit potential, as well as what prices and contracts are available. Based on that decision, the rest of the rotation falls into place for the remaining five years.

There is no single typical rotation in organic production. You are able to design and implement what works best for you. The weather and soil conditions on your farm, along with market projections, will all be important factors that guide you. However, there are some common components in organic crop rotations. In both of the above rotations, for example, small grains play a key role. You would never see this happen on a conventional corn and soybean farm.

When it comes to planting the crop, seed and variety selection play a critical, but different, role than they do in growing conventional crops. In my years as a technical college agriculture instructor, we spent a lot of time in the classroom reviewing university and company seed plot results with respect to yields and standability in corn or insect and disease resistance in small grains and alfalfa. The gold standard was yield test results published by our land-grant college, the University of Wisconsin.

Matt Dillon, one of the founding members of the Organic Seed Alliance has years of expertise in promoting plant breeding for organic crops. He points out that strategies conventional farmers use to select crop varieties often don't work very well in organic seed and variety selection. He suggests that the best strategy in soybeans, for example, is to look at varieties that are capable of fending for themselves in a hostile environment. In using this example, he points to plant traits such as the plant getting off to a quick start to get ahead of weeds or having better traits for withstanding insect attacks or perhaps a bigger root system that enables it to get more nutrients out of the soil.

You won't find this sort of information in typical university test plot data. Varieties that do well in a standard test plot trial may do poorly under organic growing conditions. Talking to your organic farming neighbors and getting information from seed companies that have organic seed lines and do some of their own breeding is one of the best ways to get the information you need in your decision-making. Another key source of information is an organic grain marketer. Oftentimes, sending a sample in advance of an agreement to purchase is part of the organic grain marketing process. Organic grain marketers, therefore, have a lot of samples come across their desk. Because of this, the marketers will have a good idea of what varieties do well under different market conditions.

Fertility strategies to insure adequate nutrients for the crop you intend to grow are critical. Nitrogen sources are the most limiting factor in soil fertility. Thus the emphasis on legumes is important. Your land grant university college of agriculture, extension agent, or agronomist will have information on the available nitrogen credits you can calculate into your fertility needs from plowing down alfalfa, soybeans, or other legumes. There is also high value in having a livestock component for manure in organic production. One of the farmers I mentioned earlier uses liquid hog manure from a neighboring farm, and the other uses manure from a cow-calf herd on the farm. I have also been on organic farms that used composted chicken and feedlot manure. However, manure production from animals on the farm has to be balanced against having too many mouths to feed and therefore cutting into profits from cash crop production. Some liquids for foliar feeding have been approved for use by organic farmers, but they don't tend to have high nitrogen content.

You must also choose your crop rotation with an eye toward controlling weeds, insects, and plant diseases. A good rotation allows nature to intervene and break cycles that promote these problems. In any rotation and with any crop, how you control weeds is critical.

Legume crops control weeds nicely once they are established. Seed bed preparation, planting in critical windows, and mechanical weed control methods such as rotary hoeing and cultivation will all be important to weed control at different stages in an organic rotation.

Timeliness in organic farming is very different than for conventional row crops. In conventional row crops such as corn and soybeans, the minute the weather shows signs of warming up in spring the race is on even if the soil temperature is colder than optimal for germination. On the other hand, timeliness in organic farming means having the patience to wait for the right soil temperatures, doing tillage to maximize weed control along with timely rotary hoeing and cultivation. It also means having the patience for the crop to dry down to optimal harvest moisture.

Smalls grains, especially oats, barley and field peas, do well in early cooler weather. On the other hand, most organic corn is planted later (usually about two weeks later, depending on weather) than conventional corn for a number of reasons. The soil is worked initially to get control of the first weed growth. Later planting allows the soil to warm up and for the corn to get a head start on the weeds. It also allows for tillage to get a second growth of weeds before planting. Then comes timely rotary hoeing (usually two times) followed by cultivation twice. A second reason for later planting is that in a normal growing year, your neighbors' GMO corn will be past pollination when your corn is pollinating. This reduces the chances of GMO pollen contaminating your organic corn, which can lead to rejected loads when it's marketed. Like corn, soybeans are also always planted later so the soil can warm up and so that more thorough weed control methods can be used. Soybeans can also be planted later because they have a shorter maturity than corn.

The more rotations you see on organic farms, the more you see that about six crops tend to receive the most attention. Those crops are corn, wheat, soybeans, alfalfa, oats. and barley. Oats and barley are used as a nurse crop for alfalfa. Other legumes, such as any of the clovers, can be used as a source of nitrogen, but they tend to have a shorter life of one to two years. Another emerging market trend in organic field crop production is feed rations, especially for poultry, that have no corn or soy as part of the ration. This has created a market for other grains such as field peas as a protein source.

Some of the more common grains, their description and characteristics for marketing are:

Corn-Food Grade: Follows the USDA's grading system for #1 yellow. Key market characteristics include good test weight, color, and kernel size. Corn that goes for food grade usually is tested for GMO presence to meet strict European standards.

Corn-Feed Grade: Closer to #2 yellow corn, organic feed grade corn is most commonly used in dairy and poultry rations. Prices will vary based on test weight, which is a comparison to the standard 56 pound bushel weight.

Soybeans-Food Grade: Soybeans that fall into this category tend to be variety-specific for markets. Some of the more common varieties are Vinton 81, HP204, Iowa types, clear hylum and, occasionally, black hylum. Beans that are sold into the food market must be whole and not shattered or damaged during field harvest or handing. They must also be free from dirt, stains, and stones. As with food grade corn, testing for GMO presence is almost always done.

Soybeans-Feed Grade: Beans that fall into this category often are specifically raised for feed. Beans that fail to meet food grade criteria will also fall into this category. The most common descriptor is #2 beans or screenings which include split beans. Beans for feed grade generally sell for a lower price than those grown for the food market. Some type of processing is required to make the best feed value for livestock. Dairy and poultry rations are big users of soybeans processed into meal. Another possibility is roasting the whole beans.

Barley-Milling Grade: The standard weight for a bushel is 48 pounds. Good color, kernel size, and test weight are important.

Barley-Feed Grade: Barley in this grade can be used in feed rations at 80% of the nutrient value of corn.

Hard Red Winter Wheat: This is the dominant class of wheat grown in the United States and is the most commonly exported wheat. It is used in bread, rolls, and to some extent sweet goods and all-purpose flour. Wheat in this market category would grade #1 according to USDA standards and have about 12% protein and good test weight. Different millers have different protein requirements depending on how their end users use the product. Some millers may blend higher or lower protein wheat to get the specific protein content they need. Manufacturers of products such as pasta and bakery flour also factor in what are called "falling numbers." Falling numbers measure the effect of the starch-degrading enzyme alpha-amylase on wheat. Sprout damage is bad for bread quality. A falling number above 300 indicates no sprout damage in a wheat sample. It is best to communicate with your buyer or marketing agent to understand how falling numbers used in wheat or rye might affect the grain you have to sell.

Hard Red Spring Wheat: This is grown mainly in the western upper Midwest including North Dakota, Minnesota, South Dakota, and Montana. To meet market specifications here, the wheat needs to be USDA #1. Test weight, protein content, and falling numbers are important indicators of what you can expect to be paid for what you grow. Yields are often less than for hard red winter wheat. Again, millers have different requirements depending on the end user. It tends to be higher in protein then Hard Red Winter.

Soft Red Winter Wheat and Low Protein Hard Red Winter Wheat: Should grade USDA #1. It tends to be grown mainly east of the Mississippi River. It is used for flat breads, cakes and crackers. It can be hard to sell in the food market if the protein is 8% or below.

Soft White Wheat is mainly grown in the Pacific Northwest with similar uses as Soft Red. In addition it can be used in cookies, pastries, muffins and snack foods.

Hard White Wheat is milder and sweeter in flavor and used in yeast breads, hard rolls, tortillas and oriental noodles.

Durum Wheat is grown mostly in North Dakota for use in semolina flour and pasta.

Spelt is often referred to as an ancient wheat cited in the Old Testament. It is grown on occasion in lower Michigan and in Ontario, Canada. It is fall planted and its cultural practices are similar to winter wheat. It is in demand in health food stores as flour. Its nutritional value tends to be similar to oats. Organic farmers and marketers I have talked to note that it is best to have a contract in hand before even considering planting spelt. In some years, supply exceeds the demand and hold-over of a crop for a year or two is not uncommon.

Feed Grade Wheat: Any of the wheat mentioned above can be sold for feed, especially if it is 11% protein or below and doesn't meet miller specifications. In some years of large amounts of rain and high humidity, vom-itoxin can make the wheat unsuitable for human consumption. Feed wheat can be used in dairy and poultry rations.

Other crops: Organic hay (alfalfa and combinations with grass) is sold on the basis of tests and Relative Feed Value (RFV) and Relative Forage Quality (RFQ). Additional organic crops include oats (food and feed), buckwheat, dry edible beans, flax (golden and brown), millet, rye, field peas and triticale. Growing for very specialized markets like white wheat, blue corn, and white corn usually requires a contract and specialized seed varieties to meet the buyer's needs.

Ultimately, first and foremost it is always important to investigate markets and to explore options before deciding to grow some of these crops, especially millet, flax, triticale, dry edible beans, spelt, and buckwheat. This is where having a farmer-oriented marketing agent can and should be a key part in helping you decide what will work for you. He or she can also anticipate what you will have to sell and can look for opportunities to contract the crop at favorable terms to you before the crop is harvested.

Developing a rotation that balances economics and biology is the very heart of organic farming. The time you spend in long-term planning will, as much as anything else you do, determine your success as an organic grain farmer.

CHAPTER FOUR

Contracting Basics

The organic grain market is nothing like what you see in conventional grains. With conventional grains, futures prices and local prices are always clearly spelled out. You can't do anything about them, but there they are. Your only decision is whether to sell now or later. The USDA regularly collects market information and reports it for various major grains on a daily basis. There are no such well-established prices for organic grains. The USDA only has been reporting organic grain prices within the last several years. The reports are issued every two weeks. When organic grain markets are changing, these reports can become quickly outdated.

In organic grain markets, you need to find a buyer, settle on a price, and negotiate other important matters. Finding buyers, knowing what to ask, reliable price discovery, and all the related issues that come up are big challenges for any organic grain farmer, especially for one just getting started. Carmen Fernholz farms about 400 acres in western Minnesota. According to Carmen, you can figure spending 15 to 20 percent of your time each week on marketing what you have to sell if you choose to do it yourself. There are those who thrive on this challenge and others like Carmen who turn it over to a marketer or agent.

Whether you do it yourself or use a marketer, nothing is ever done in organic grain marketing without some type of contract. Contracts are something to understand and work with and, if done correctly, nothing to fear. When both parties understand their responsibilities in a contract, things generally go well or can be corrected when they don't. But things can get expensive in a hurry when one or both parties don't fully understand what a contract requires.

A good primer on organic grain contracting is the *Farmers' Guide to Organic Contracts* published by the Farmers' Legal Action Group (www.flaginc.org). The Guide lays out "Seven Rules of Contracting" which are important to keep in mind. I want to go through those rules here and tell you why they are especially important in marketing organic grain.

Rule 1. The contract almost always favors the party who wrote it.

In most cases, the buyer writes the contract. At one time or another, most of us have gotten loans from a bank, credit union, or farm lending agency. A good share of the time, we farmers are trusting people and we do business with those we like and get along with. When is the last time you actually took home all the loan documents and carefully read them before taking them back to the bank for signing?

In my years of teaching farm business management to adult farmers, I often used this simple example. What does your banker, who may be the best friend in the world, see from his or her perspective when you walk in the door to discuss a loan for a car or your farm? It comes down to something very simple. If I lend you the money, I'm taking a huge risk as to whether I will get the bank shareholders' money back. What are the prospects in a worst-case scenario that I will have to sell you out to get back the money I loaned you? Will I have to repossess a car or machinery? How much time, expense, and bother will it be? There are generally whole paragraphs spelling out options for the lender in case of default. Language that protects you as a borrower will be much harder to find.

Organic grain contracts are no different with regard to attention to detail, especially if they are written by the buyer and simply handed to you to sign. You may be forced to change your crop rotation or other ways that you farm to meet the contract obligations. On the other hand, volume talks when it comes to contract negotiations. In many cases, because of the volume they handle, cooperatives are in a much better position to negotiate terms of contracts favorable to their members than an individual can get.

Rule 2. The buyer can force you to fully perform your contract promises.

Suppose you sign a contract to deliver 10,000 bushels of corn. What happens if you have a short crop and only come up with 8,000 bushels? With a written contract, the buyer could force you to go out and buy and have delivered the 2,000 bushels you are short. There is also a good chance that if your crop is short, you aren't the only one with substandard yields and prices could rise substantially for the corn you have to buy. In addition, the buyer would expect the quality standards of the contact to be met. It is better to err on the safe side when it comes to contracts by contracting a portion of your crop or to have a number of contracts to spread your marketing throughout the year.

I was once part of a team coordinating a fairly large organic grain contract with a major buyer. At the time the contract was initiated, the prevailing organic market was between $7 and $7.50 for feed grade corn. The contract price was for about $8.00, so the agreed-upon price seemed like a good one at the time. Producers were advised to only commit a part of their crop to the contract and wait to see what happened to the markets. Because it was such a good price, one farmer, against everyone's best advice, insisted on committing his entire crop. When it came time for delivery, market prices had risen way past $8.50 on towards $9.00 and $10.00. The producer tried every which way to get out of delivering the crop at $8.00. In the end, after being threatened with legal action, he delivered reluctantly. If he hadn't delivered, he would have had to hire an attorney and would have almost certainly lost in court. Then he would have to pay not only his legal fees, but those of the buyer to collect the money. The cost would have been much more than simply taking the $8.00 per bushel which was a good price at the time he made the commitment.

Rule 3. You cannot assume your failure to perform contract promises will be excused.

Back in the 1990's, conventional grain farmers across the U.S. were sold on the idea of really cashing in on a new type of contract proposed by the grain industry. I will spare you the details of how "hedge to arrive" worked. When market prices moved in the opposite direction from what farmers who signed the contracts thought they would, they were sued for millions and millions of dollars for failure to live up to the terms of the contracts they had signed. Thousands of farmers took a financial beating and some of them even went bankrupt. Assuming that contract problems will simply "go away" didn't work for them, and it won't work for you, either.

For example, if you contract 10,000 bushels of corn, but deliver either more or less, the buyer may be in an understanding mood. But don't count on it or assume that will happen. Many organic farmers forward contract a portion of their crop and then lock in additional contracts and prices after the crop is in the bin as a known quantity. This brings me to another advantage of group marketing—an experienced agent, representing many organic grain farmers, can often negotiate with the buyer to fill the terms of the contract with other farmers' grain. In addition, that marketer can often move your grain to a buyer and market that more closely fits the grain you have to sell.

Rule 4. Oral promises made during negotiations or after a contract is signed mean nothing if they are not written into the contract.

Even if a contract is already signed, it can be amended to correct misunderstandings or better reflect current conditions. But whether you are looking at a new contract or amending one already in place, what is mutually agreed upon needs to be put in writing and spelled out clearly. This rule seems simple and obvious. Don't be fooled, however. Too many times, I have seen taking things for granted in a contract lead to big trouble later.

For example, I recall a situation in which some producers had contracts with a buyer that stipulated strong prices. As time went by, prices fell below what the buyer had agreed to pay in the written contracts. The buyer would agree verbally to take delivery of a number of loads under the contract for the next week. Then, after loading and delivery with a trucking firm had been scheduled, the buyer would call back at the very last minute and cancel the deliveries. The buyer used such delaying tactics to source cheaper grain rather than pay according to terms of the contract.

This is but one of many examples of verbal promises that were made and then not followed up on to evade the terms of the original contract. Remember, the basic lesson is, "Don't just take the buyer's word for it." Putting it in writing means there is less chance of a misunderstanding and more options for you if things go wrong.

Rule 5. Language that sounds "reasonable" or that the buyer describes as "standard" might have hidden consequences.

Suppose you contract grain for delivery in December under "standard" terms. Depending on where you are in the Midwest, this might be "reasonable" and you have more time to do things like load trucks. That doesn't mean that things like weather and the holidays don't come into play. Weather could mean a major snowstorm hindering or slowing trucking and oftentimes there are reduced workforces at that time of the year. Such circumstances can delay or hinder both your ability to ship the crop and your buyer's ability to accept delivery. And about the time you need that check for the grain you sold, the key office people could decide to take off the entire week between Christmas and New Year's.

Another guessing game in marketing is road weight restrictions, especially in the Upper Midwest. Unpredictable weather could find you facing having to ship grain with weight restrictions slapped on roads at a moment's notice, especially towards spring. This could end up costing you more money by having to ship lighter loads or not being able to get them out at all.

I have also seen situations in which a producer has a contract that has a good price, but the market falls dramatically prior to delivery and payment. Under these circumstances, a buyer may be reluctant to take your high-priced grain when he or she can readily source lower-priced grain. How do you handle it when a buyer wants to either cancel your contract or is slow to take the grain and very slow to pay? Once again, an experienced marketing agent, representing large volumes of grain, might come in handy.

Testing for the presence of GMO's is becoming increasingly important in selling organic grains. Grains going into the food market will always be tested carefully and some, like corn and soybeans going into food products for export to the European Union, will be among the most carefully tested of all. But no matter what the use or destination, it is "standard" procedure to test for GMO presence in organic grains. There are a couple of different types of tests. One that is relatively cheap and fast is what they call a "strip test" or "rapid test." It consists of taking a representative sample of grain, crushing it, adding chemicals and putting the resulting extract on a strip that will turn colors if GMO DNA is present.

A problem with this "standard" procedure is that the strip test can sometimes give a false positive reading; that is, your grain could be clean but show contamination. Do the "standards" in your contract allow for a retest? If so, how many? In some cases a more precise test can be run, but it can cost several hundred dollars. Who pays for the test or tests? Will it be the buyer running the tests or an independent lab? Simply saying we will deal with that issue if it arises is not a good option and usually a headache for both you and the buyer. I've heard of cases where the buyer simply notified the producer and then took a load across the driveway to his or her conventional grain operation and paid the farmer the conventional grain price. This was a tremendous loss to the farmer.

Rule 6. Higher price terms might not mean larger payout.

You can contract for higher prices but take home less money because of yield loss and other factors. I see this most often with food grade corn, wheat, soybeans, barley, or oats. For example, the price for food grade soybeans can be $5.00 or even $6.00 more per bushel than feed grade beans. However, the varieties used for food grade beans tend to yield lower than other varieties used for feed grade. Food grade beans also require more careful cleaning (removal of stones, dirt, inert matter) beyond what comes off the combine and that can run up your cost. Diseases such as "blue stain" can result in discoloration of the beans and failure to meet contract terms. Many times, food grade beans are exported and buyers, especially in Asia, are very specific about the qualities of beans they want and need. Another consideration is whether the buyer and his customer will tolerate GMO presence in the grain which will invariably be tested. Rejection due to GMO presence often means finding alternative markets at a much lower price.

A farmer I know had soybeans and corn shipped to a buyer for the food market. The grain was excellent and he took every precaution to ensure complying with organic standards. No contamination occurred during shipping from his farm. Upon testing at the buyer's delivery point, however, his crop came up positive for GMO contamination and he lost over $15,000 in a matter of about two weeks on what he sold. If there was any saving grace, it was that he had an experienced marketer through his cooperative who was able to find alternative markets and soften the blow from having to simply dump it in the conventional market. Oftentimes a marketer working on behalf of farmers is the only one with enough experience to know of alternatives or has the negotiating skill when situations like this arise.

Rule 7. Contracts are subject to negotiation, and you can walk away.

Contract negotiation is a bit like playing poker—you got to know when to fold 'em. Sometimes the best thing you can do is walk away from a negotiation. An everyday example is buying a tractor, truck, car, or piece of machinery for your farm. Most of us have come very close to finishing the deal and just couldn't make it to the finish on closing the deal to our satisfaction. So we walked away and either changed plans or looked some more for a deal that better met our needs. The same is true with organic grain contracts. If your gut feeling is that it's not a good deal, it probably isn't.

One of the OFARM marketers I work with walks away from contracts that have more than five criteria for the grain to be delivered. In such cases, he looks for another buyer instead of taking the chance that at least one of the criteria will be a "deal breaker" at harvest time. He cites food grade oats as a good example. Meeting buyer specifications for food grade oats can be difficult and not worth the often small price differential over feed grade oats.

Another example comes to mind of a friend who is also an organic grain marketer. He received a call from the buyer of a load of grain from one of his producers. The load was sitting on the scale at the buyers and tested positive for GMO's. The buyer rejected the load. In the ensuing discussion, a misunderstanding had occurred from the outset between the buyer and the marketer as to whether they were talking loads of grain to be delivered or actual bushels. It was easier for the marketer to conclude that the correct bushels had been delivered and to find another market for the load of grain than to continue to negotiate.

Some Other Things that Will Help You Get the Right Contract

There are a couple of additional provisions and types of contracting for organic grain that can be very favorable to you as a producer if negotiated right.

One is for long-term contracting. Some producers I know will contract a year ahead if given the opportunity. In other words, as they are growing this year's crop they will contract for a small part of next year's crop. In doing this, they have some idea of what price they can sell for on part of next year's crop and can use that information in long-term financial planning. A key to doing something like this is contracting for acres as opposed to a specific number of bushels for delivery. In other words, the buyer agrees to take whatever bushels are produced on a set number of acres.

No grain or hay is ever sold without some sort of sampling. The sampling needs to be accurate and representative of what you are selling. Proper sampling techniques and keeping a part of the sample for further testing if a dispute arises are critical. This is usually done by sending a sample in advance to the buyer. Then, when the grain is loaded for shipment, another sample is taken and the farmer retains half of that sample for use if questions arise.

Another useful provision to have in a contract is called an "Act of God" clause. Its inclusion needs to be negotiated. For example, in case a hail storm or other weather incident wipes out your crop, you can be excused from delivery due to circumstances beyond your control. The buyer of your grain may also have or want this type of provision for his or her protection. Suppose that the delivery point for your grain has a major fire or a grain bin collapses limiting storage capabilities. If the buyer has "Act of God" protection in a contract, you must think through your "Plan B" if that buyer protection comes into play.

A Last Thought

One of the advantages of hiring an agent is having someone who is in the organic markets every day on your side of the bargaining table. A good marketer can help you develop a strategic marketing plan, think through your options, and negotiate contract terms. Some farmers excel at and like negotiating. Many do not because they don't do it every day. More often than not, the bottom line is that an experienced marketing agent working on your behalf is the way to go.

I once saw an advertisement that said, "In business, you get what you negotiate, not what you deserve." I'm sure that is true for all businesses, but none more than organic grain farming. Make sure you have the knowledge and horsepower you need to hold your own in contract negotiations with experienced buyers representing companies much larger than your farm. Your success as an organic grain farmer depends on it.

CHAPTER FIVE

Why Organic Standards Are So Important

No matter what you grow or how you grow it, there will be national and global issues that play a vital role in your success. Organic grain farmers are no exception. As I write this, we are in the middle of controversies over imports, liabilities for GMO drift, and a new organic commodity check-off system. As important as these issues are, however, none deserves your attention more than the standards determining whether what you grow is certified organic.

The early organic farmers and other concerned groups fought hard to put in place standards and mechanisms so that farmers, brokers, processors, distributors, and retailers would all play by the same rules. The Organic Foods Production Act of 1990 became law and USDA's Agricultural Marketing Service was charged with implementing provisions specified in the Act. This process resulted in the USDA's organic seal of approval. The National Organic Program (NOP) was created within the USDA to implement standards, issue rules and enforce those rules. The National Organic Standards Board (NOSB) was then set up to bring stakeholders in the organic industry together so that they could provide input to the NOP on organic standards, rules, and regulations.

As you can see, a system that can modify the organic standards is part of the process that established the standards in the first place. A formal way to keep standards current seems reasonable enough at first glance, but if you are an organic farmer, you should also see it as a major threat to your livelihood. As far as consumers go, the USDA organic seal is your "brand." The value of your brand depends on the exact wording of some important standards. Changes to those standards are always under consideration. Some of those changes can transfer value from your farm to global processors and retailers.

Grocery shelves now contain organic products from General Mills, Kraft, and many other global food processors. This makes them stakeholders in the organic industry, and therefore means they will have strong representation in the National Organic Standards Board (NOSB). A report by Food & Water Watch[1] outlined the concern in clear terms:

> *The largest food processing companies have worked to weaken the rules governing organic food. Giant traditional food manufacturers and agribusinesses with valuable organic lines (like General Mills,*

---

[1] Food & Water Watch. *The Economic Cost of Food Monopolies.* Washington, DC. November 2012.

Campbell's Soup and Driscoll Strawberry Associates)
have had company representatives on the USDA ad-
visory board that establishes the standards for or-
ganic farming and food manufacturing, and over the
past decade the number of approved non-organic
substances allowed in organic food has jumped from
77 in 2002 to more than 250 today. Once standards
are put in place, USDA's lackluster enforcement fur-
ther dilutes the organic label.

You should anticipate threats to your organic brand, and therefore to your livelihood, in two important ways: (1) direct changes in the standards, and (2) creation and promotion of confusing alternative brands such as "natural" and "GMO free." Organic grain farmers are especially vulnerable to both types of threats.

As weather patterns disrupt organic grain production, it is easier for corporate livestock producers to argue that we need relaxed standards to cope with organic feed shortages and higher organic feed costs. The seriousness of this threat cannot be stressed enough, for without the standard requiring that certified organic livestock consume only certified organic feed, the market for most organic grain farmers could largely disappear. Relaxing the feed standard has been tried once, and no doubt the issue will come up again.

In 2003, an organic poultry entity in the southeastern United States complained that they couldn't get adequate supplies of certified organic grain to feed their chickens. They successfully inserted a provision in the Farm Bill that would "relax" the organic feed standard under certain conditions. The *Los Angeles Times*[2] told the story this way:

> *With a day's handiwork, a Georgia Republican looks to have undone what it took organic farmers and environmentalists more than a decade to achieve. A last-minute provision inserted in the 2003 federal spending bill at the behest of Rep. Nathan Deal would loosen restrictions in the federal Organic Standards Act and allow poultry farmers and other livestock producers to avoid using organic feed if it costs more than twice as much as conventional feed grown with pesticides. Yet they'd still get to label their products "organic."*

The entire organic community came together and got legislation passed repealing the provision. However, it was a seven-month process to get the repeal enacted into law. I'll leave it to you do decide what your market would be like if that law were still in place.

The second threat facing organic grain farmers is that of alternative brands being promoted by processors and retailers. What happened when Dean Foods bought Whitewave, the owner of the Silk brand of soy "milk," is a lesson none of us should forget. The story be-

---

[2] "Bill may loosen organic standards." *Los Angeles Times.* February 14, 2003.

gins when Dean used its new ownership of Whitewave to begin negotiations on what the company would pay for the organic soybeans that went into Silk soymilk. Oren Holle, a good friend and organic grain farmer in Kansas, was at those negotiations in his capacity as President of Kansas Organic Producers.

Dean Foods was looking to find a larger supply of useable organic soybean varieties so it could expand production. Oren Holle was among the farmers who proposed a partnership between Kansas Organic Producers, Midwest Organic Farmers Cooperative, and an organic seed supplier to determine where those beans could best be grown and which varieties to plant. A meeting was organized to formalize a business relationship and to discuss, among other things, a pricing structure to make the arrangement work for both Dean Foods and the farmers.

At the time, it appeared organic growers would be willing to involve themselves at a profitable price of $12.50 per bushel at the farm gate. Dean Foods quickly made it clear that it could import beans for $11.00 to $11.50 . If the growers were not willing to do the deal at that price, the arrangement could not work and imported beans would be used. Just like that, the deal was over. Dean Foods promptly reformulated with conventional beans, some from China, and then succumbed to consumer pressure to go with a GMO Free label. They never got back to using organic soybeans, however.

This was very costly for organic soybean producers. According to an analysis by Food & Water Watch[3], the switch to so-called "natural" soy milk cost organic soybean farmers 1.2 million bushels per year in lost demand. Something like $8.7 million per year went out of growers' pockets and into those of Dean Food investors. How could Dean get by with such shenanigans? Why didn't its rivals blast them for what they did and tout their own organic products? The answer is the same answer we have for so many of our economic problems today: corporate concentration. Dean had such a large market share in soy beverages that there were no other companies to effectively counterbalance what Dean was doing.

These examples, and many more we either have seen or will see, point out a fundamental difference between marketing conventional grain and marketing organic grain. Both have standards, but the standards for organic products play a far more important role. Those standards are the organic farmer's "brand" and, to a significant degree, set the value at which organic products can be sold. Because of this, organic grain farmers need better communications with those who consume their products and a strong voice at USDA. No individual organic farmer is large enough or powerful enough to do this job on his or her own. Only organized farmers can do that.

---

[3] Food & Water Watch. *The Economic Cost of Food Monopolies*. Washington, DC. November 2012.

CHAPTER SIX

What's Your Story?

There are more certified acres planted to organic feed corn in the United States than are planted to all organic vegetables combined. Farmers like you plant more acres of organic soybeans than all of the organic fruit growers combined. Depending on feed availability, thousands more acres of the wheat, oats, barley, sorghum, and other grain crops you grow are also fed to livestock. The biggest single use of organic cropland in the United States is hay and silage. Most of you will grow hay and possibly silage as part of the crop rotation systems that improve soil and protect the environment.

In spite of the wonderful environmental advantages provided by organic grain farming, most consumers have never met even one of you. Some of what you grow is used to make bread and soy beverages, but most of it goes to feed organic chickens, cows, and pigs. Because of this, you are not the farmers the public typically meets at a farmer's market or sees handing out samples at the local co-op or grocery store.

No one is likely to tell your story for you. Even if they did, I doubt they would do as good a job as you can. But what is your story? Why is the way you farm so important for the well-being of everyone, not just you? Is it environmental benefits, benefits to rural communities, or maybe a combination of these and other factors?

On the following pages, you will see how some other organic grain farmers have told their stories through a project I led. You will see that when consumers buy organic dairy and meat products, they become part of a special partnership with organic grain farmers to protect our nation's farmland. You will see how that partnership has wonderful environmental advantages for water quality, for biodiversity, for bees, and for human safety. None of these advantages are guaranteed by half-way labels like "natural" and "GMO free."

Use these stories from other farmers to think about how best to tell your own story. Then get out and tell that story wherever and however you can. The future of organic farming is in your hands!

Because I grow feed crops organically, the soil biology on my farm is healthier. This, in turn, increases the nutrient density in the eggs my chickens lay.

The more Larry Heitkamp learned about soil health, the more he came to realize that chickens were an integral part of the circle of life on his organic farm.

Heitkamp sells organic eggs from the chickens he raises on his central Minnesota farm. "My chickens are free range and forage on grasses all summer long," says Heitkamp.

He also grows organic oats, barley, and peas to make sure the chickens are well-fed in the winter months when grass is not available. He takes his crops to an organic feed mill in nearby Wadena, Minnesota, where they are ground and mixed into a balanced winter diet for his chickens.

The way Heitkamp raises chickens is nothing like the way eggs are produced on industrial farms. A visitor to his farm will find no giant confinement buildings in which thousands upon thousands of birds live miserable lives. No semi-trucks bring chemically-grown GMO feeds in from distant sources. No antibiotics find their way into his feed mixes.

The organic difference translates into healthier soils, more humane treatment of animals, and a better, safer product for consumers. "Because I grow feed crops organically, with cover crops instead of fertilizers and chemicals, the soil biology is healthier," says Heitkamp. "This, in turn, increases the nutrient density in the eggs my chickens lay."

Larry Heitkamp sees his customers as partners in a grand plan to replenish the land and raise healthy families. He puts it this way: "When you buy my organic eggs, you feed your family and heal the earth."

The only way I know to solve this bee problem is to not be using the chemicals, especially the farm chemicals, that we now use. That's why I'm glad I'm an organic farmer.

Insect-pollinated plants provide us with one third of our food supply. Yet the honey bees and wild bees farmers so depend on are dying at an alarming rate. What can we do to save the bees? According to Tom Bilek, the answer is simple: farm organically.

Bilek has been an organic farmer near Wadena, Minnesota, since 1998. He knows first-hand how important bees are to farming success. Bees pollinate many of the organic crops he grows on his farm, especially the hay crops that feed his cattle.

Bilek also knows a simple, effective way to keep bees safe and thriving. Farm chemicals are often cited as a leading cause of decline in bee populations. Because he is an organic farmer, none of those chemicals are ever used on his farm.

"The only way I know to solve this bee problem is to not be using the chemicals, especially the farm chemicals, that we now use," says Bilek.

Farm chemicals are not the only problem for bees. The oceans of corn and soybeans that dominate the industrial farming landscape are a terrible habitat for bees. There are no nourishing flowers to provide bees with protein from pollen and carbohydrates from nectar.

Here again, organic farming is the obvious way to avoid this problem. Instead of planting only corn and soybeans, year after year, Bilek plants a variety of crops that lead to healthier soil and greater biodiversity. "Bees love the buckwheat I grow on my farm," says Bilek.

Tom Bilek knows that bees do very well on organic acres. "That's why I'm glad I'm an organic farmer," he says. "And that's why I'm glad you buy organic products. We're working together to save the bees, organic farmers and organic consumers."

What's the best way to keep toxic chemicals and synthetic fertilizers from harming water quality? Don't use them in the first place!

Countless studies remind us that farm chemicals are bad for water quality. Jackie Keller, a grain and livestock farmer near Topeka, Kansas, doesn't take chances with the water on her farm. She converted to organic agriculture.

"As an organic farmer, I don't use any chemicals that can harm water quality. That's the best way to avoid polluting chemicals—don't use them in the first place," says Keller.

Her efforts have not gone unnoticed. Keller won Shawnee County's 2011 Water Quality Award. She was lauded for her "impressive step further on her Fully Certified Organic Farm" where "zero petro-chemical pesticide use means zero pesticides leaving the fields and entering the creek."

Keller has the same issues with weeds as her industrial farming neighbors. But instead of using toxic chemicals to kill them, she uses mechanical equipment to hoe and cut down the weeds. "The cattle I keep on my farm also do a good job of eating weeds," says Keller.

Synthetic fertilizers can also pose a threat to clean water, but not on Keller's organic farm. Instead of planting the same crops on the same ground year after year, she rotates crops that need nutrients with those that produce nutrients naturally. Her crop rotation also prevents erosion by keeping the ground covered with soil-preserving grasses.

When Keller needs additional nutrients for her crops, she uses livestock manure instead of synthetic fertilizers. "I prevent nutrients from running off into streams by working manure into the soil as soon as I apply it," says Keller.

"My conservation officer told me that it is 200 times more expensive to clean up polluted water than to prevent pollution in the first place," said Keller. "Organic farming is just common sense when you look at it that way."

Jackie Keller, like thousands of other organic farmers, works hard to provide safe food and use safe farming methods. Every time you shop for the USDA organic label, you support their efforts to replace industrial farming with a farming system we can all live with.

To be organic is to be pro-people, pro-environment, pro-consumer, and pro-community. When you purchase organic, you are also purchasing all those virtues and all those values.

Charlie Johnson has been an organic grain farmer near Madison, South Dakota, since 1976. After all these years, his guiding principle is still something from his late father who first farmed the land organically.

Johnson's father Bernard said, "Anything that is going to go on our land must be something that can also go on the tip of my tongue." In other words, anything that was to go on the Johnson's land could not cause harm to anyone working on that land.

His father's wisdom is paying off. "We have four seasons of wildlife on our farm. Be it deer, pheasants, ducks, or bees, they all find a safe harbor here," says Johnson. "Everything thrives and does well here on our farm."

"We believe in diversity, we believe in a safe environment, and we believe in a healthy environment," says Johnson. Organic farming is a perfect match for his values.

Aside from avoiding all toxic chemicals that could hurt people and wildlife, biodiversity comes naturally in his farming system. Synthetic fertilizers are replaced by planting a series of different crops that provide nutrients in a natural way. The resulting diversity of plants on Johnson's farm provides much better wildlife habitat than the borderline monoculture that has become standard operating procedure for industrial agriculture.

Charlie Johnson sees himself as farming in partnership with organic consumers. Both Johnson and those who eat what he produces share the common goals of nutrition, quality food, conservation, and a healthy environment.

"To be organic is to be pro-people, pro-environment, pro-consumer, and pro-community," says Johnson. "When you purchase organic, you are also purchasing all those virtues and values."

When you buy organic, you support farmers who are working hard to feed the world in ways that keep our planet safe and healthy.

"Many people think organic farmers are low-tech," says Carmen Fernholz. "That may be true for some, but the organic farmers I know use the same hi-tech equipment you would find on any farm around here, conventional or organic."

Fernholz harvests corn, soybeans, and other grains on his southwestern Minnesota farm with the same type of machinery that his conventional-farming neighbors use. With one pass through the field, he can harvest grain, monitor yields on his on-board computer system, and empty the crop into a truck that takes it to a storage facility on his farm. Fernholz also uses the same type of tractor found on any farm in his part of the state. It's equipped with state-of-the-art GPS navigation equipment that guarantees straight rows and accurate field work.

But Fernholz draws the line at using equipment that doesn't suit organic production. "Some of my farm machinery, while still hi-tech, is specialized for organics," says Fernholz. Controlling weeds is a good example. "On my farm you won't find a giant spray rig or any of the toxic chemicals those spray rigs apply to control weeds," he said. "I use separate machines for different weed problems, in much the same way other farmers would use different chemicals. But the methods I use are safer for the environment and my family."

Carmen Fernholz is adamant that using appropriate modern technology in organic farming is the way of the future. "The next time someone tells you organic farming can't feed the world because it is primitive and low-tech, show them some pictures of how I farm. More often than not, they will have a hard time telling the difference between the technology I use and that used by any other farm in this part of the state."

There is a difference, however. The difference is in the label Certified Organic. It assures the best food for consumers, the best practices to take care of the countryside, and the promise of a world fed in the most sustainable way possible.

It's a pretty neat thing when you can go from field to bottle, all within a two-mile radius.

Paul Graham is the president of Central Waters Brewery in Amherst, Wisconsin. Central Waters Brewing Company services over 200 retail locations and has expanded its product line to include 18 beer styles.

For the past six years, he has bought organic barley from Bob Stuczynski, a nearby organic grain farmer. "Bob's barley is a quality product," says Graham. "And, of course, being organic is a big thing with us. We use some organic grain in every single beer that we make."

Being organic is also a big thing for farmer Stuczynski. "We have fields right next to our house," says Stuczynski. "So when you have small children like we do, that's the last place you want to have farm chemicals. We sleep a lot easier knowing that organic farming is safe for our family."

Graham also puts a high value on buying local. "It's a pretty neat thing when you can go from field to bottle, all within a two-mile radius," he said.

Not every organic grain farmer can be so lucky as to have customers close by. But because they are organic, they all have some important advantages for local economies that industrial farming does not have. Organic grain farmers don't buy chemicals and synthetic fertilizers from distant sources, nor do they pay the fees global agribusiness giants charge for using GMO seeds.

The Stuczynski farm is a good example. Rotating different crops and using cover crops provides nutrients and breaks up the life cycle of weeds and pests. He relies on local labor and doesn't spend any money to support distant GMO seed monopolies.

As he prepared to harvest another year's crop of organic barley, Stuczynski said, "Harvest time is a good time to remember that organic farming is not just good for our family. It's good for your family, too."

I'm grateful organic farming has given me the opportunity to stay on the land.

Harold Wilken, who started out years ago as an industrial farmer, has never regretted his move to organic practices. "I've been very blessed to be organic, and I don't see myself ever going back to conventional."

He and his son Ross Wilken own and operate an organic grain farm in Iroquois County, Illinois, about an hour and a half south of Chicago.

"One of the biggest benefits of organic farming has been the opportunity to bring my son Ross back to farm with me," Harold Wilken said. "There are economic advantages to the way we farm, to be sure. But there is another, even bigger, benefit as far as I'm concerned. Ross will never have to handle a pound of insecticide or herbicide, and that's very important to me."

Ross Wilken has seen first-hand how few opportunities industrial agriculture provides for young people to return as full-time farmers. "A lot of my friends from college are not having the opportunities to go back to the family farm because it is just not economically viable for them," said Ross Wilken. "I'm grateful that organic farming has given me the opportunity to stay on the land."

Both Ross Wilken and his father work hard to keep up with the latest technology. For example, a rotary hoe they built allows them to weed hundreds of acres in a single day without using any toxic herbicides. "We're organic, but we are not low-tech," said Ross Wilken.

"That's one thing I got out of my college days. You have to keep up no matter how you farm. Hungry people around the world are depending on us."

Ross and Harold Wilken don't take chances when it comes to taking care of their families and the land they farm. "Consumers should do the same," says Harold Wilken. "Play it safe. Buy organic."

Don't be misled by half-way labels like "GMO free" and "natural." Get the one label that lets you have it all: USDA Certified Organic.

"Genetically modified crops are a big problem for all of us organic farmers," says Oren Holle, a diversified organic grain and livestock farmer from Bremen, Kansas. "Most farmers around here plant GMO crops. They can contaminate what we organic farmers grow. If that happens, we can't sell our crops as being organic. We lose money, and agribusiness wants *us* to pay for the damages."

You would think that Holle would be happy to see "Certified GMO free" labels springing up in grocery stores. But you would be wrong to think that way. "A GMO Free label is certainly better than nothing," says Holle, "but it is a half-way label that misleads consumers."

When Holle attended a meeting of farmers learning how to meet the requirements for GMO free labelling, all he heard was talk of which chemicals work best on GMO free farms. "They weren't talking about farming without chemicals," he said, "only different ones that could be just as toxic as the chemicals they replace."

"The USDA certified organic label is far better than a GMO free label", says Holle. "Organic farmers don't use harsh chemicals to control weeds and insects. We don't use commercial fertilizers that can pollute our water. We plant cover crops that make sure soil erosion isn't a problem. We provide safe havens for birds, bees, and wildlife. The list goes on, but you get the point."

None of these other advantages of organic farming

is guaranteed by the GMO free label alone. GMO free crops might well have been grown with all of the chemicals and synthetic fertilizers that any other industrial farm would use.

Oren Holle says that organic food is one of the rare instances where you really can "have it all," just like the commercials so often say. "When you look for the USDA organic label, you help stop the spread of GMO's and get all of the environmental and health benefits of organics as a bonus."

CHAPTER SEVEN

Finding Your Marketing Agent

You, like every other producer of organic grains, want to be known as a reliable, dependable supplier of quality grains and forages. You want to be treated fairly and avoid costly mistakes. As veteran organic farmer Carmen Fernholz says, "If I were a buyer of organic or raw products, my job description would tell me to purchase products at the lowest possible price." Carmen wants to spend his time growing organic crops, not dealing with the "hassle factor" of marketing them. And he wants a marketer on *his* side, not the buyer's. You should, too.

In marketing organic grain, you have to be contractible, have volume and be able to protect yourself. You have to be willing to put your name on the line with a certain set of terms that have to do with getting the product marketed. You need to have a voice in the terms of the contract rather than simply accepting whatever terms the buyer first offers. Last, you have to position yourself to become a dependable supplier. It can be next to impossible to do this on your own unless you have a large volume. Some farmers excel at and like negotiating, but most of the producers I know will do better if they are represented by an experienced marketing agent.

Let me give you an example of why I say that. One of the biggest mistakes individual producers make in selling for themselves is failing to negotiate for their costs of marketing. The buyer factors in his or her costs of buying, but the farmer coming from the conventional world is not used to that. He or she pays careful attention to production costs, but doesn't have to spend as much time marketing. When that farmer goes organic, it is easy to maintain this mindset and not include marketing time when he or she negotiates a selling price for grain.

Not including your marketing time in sales price estimates can be very costly. After looking at the cost of marketing through a cooperative, the operator of a very large organic farm I know of decided to save money by marketing on his own. He did a good job of figuring production costs, but didn't factor in any marketing costs. He left over $50,000 on the negotiating table that the buyer promptly pocketed. An experienced marketing agent would never make such a mistake. In this case, the farmer would have come out ahead by paying a marketing fee and getting a higher price to cover it.

Here's another example of something you might miss if you are doing your own marketing. Payment terms should be spelled out in any contract. There have been stories of buyers holding on to producer's payments for 60, 90 or even 120 days or more. If that happens, you might not have many options because the buyer already has your grain in hand. In one case I remember, a feedlot raising organic beef acquired $35,000 of organic grain from farmers and then declared bankruptcy. A farmer I know had about 40 percent of all the grain in that sale. Fortunately for him, he had a marketer who quickly stepped in with an accountant and a lawyer. They negotiated on behalf of the farmer and recovered all of his money. I doubt the story would've had such a happy ending had the farmer been working on his own.

Selling organic grains requires careful attention to logistics and trucking. Will the buyer send a truck, or will you or your marketer have to provide the transportation? When the truck arrives on your farm, especially for food grade quality grains, you need to take responsibility for making sure the truck is thoroughly washed or air cleaned prior to loading. You simply can't take the risk of contamination from a previous load that may have been conventional GMO grain. Again, most individual farmers I know don't have time to pay that much attention to trucking—they need an agent.

I sometimes hear the excuse, "Agents are only for very small farmers." Charlie Johnson of Johnson Farms near Madison, South Dakota, works with other family members to farm 2,800 acres organically. The farm has been organic since Charlie's father started farming back in the days when "organic" was a new and strange word. Even though Johnson Farms is on the larger end of the organic farming scale, Charlie still feels he needs a marketing agent. He once told me, "If Michael Jordan can have an agent, then I can have one too."

Oren Holle, a diversified organic grain and livestock farmer in Kansas, sums it up this way, "When we look at agricultural products that a producer has spent an entire year growing, anybody can sell that product. There is a buyer out there at some price. If you want to begin to market that product you have to go a number of steps beyond just selling it somewhere." Oren, like Charlie, doesn't want to rely on what a buyer says his crops are worth. He wants an experienced agent working for him every time he goes to market.

Most likely, you are no different from Oren or Charlie when it comes to marketing what you grow. You want to spend your time making sure you grow top quality crops and don't have the time or the inclination to follow markets, learn about buyers, and arrange shipping. You want a skilled, experienced person to do that. That person, just like anyone else you hire to work for you, is not interested in working for free. Some build a fee in as a hidden cost, and others make money by selling your products at lowball prices and taking a cut from the buyer. The bottom line is that you should expect to pay marketing costs of 5½ to 6½ percent of gross sales to a person you can trust to be on *your* side, not the buyer's.

Now let's turn to the question of finding a marketer to represent you. Very few organic farms are large enough to hire a full time marketer. Instead, you will need the services of one who works for many other farmers like you. Years ago, we didn't have the farmer cooperatives that could hire the marketers. We've come a long way since then. Now we have individual cooperatives you can join to hire the marketers and an umbrella organization that allows the marketers to work together in ways that benefit you.

That organization is called the Organic Farmers' Agency for Relationship Marketing, or simply OFARM. I'd like to tell you a little about how OFARM got started, and what role I played in making sure you would have a marketing agent ready to work for you.

In the mid-90's I wasn't into organic agriculture. I did know a fair amount about sustainable agriculture beginning about 1986, but back then we didn't even have organic standards—they came in 1990. In 1997, I was working for Gene Paul, a farmer from Delevan, Minnesota, who was serving as President of the National Farmers Organization. Gene asked me to work on setting up the non-profit Institute for Rural America. I served as its first executive director and helped to seek grants for agricultural projects that would enhance the well-being of farmers and communities in rural America.

I met some organic farmers and learned of the trouble they were having with low prices for organic food grade soybeans. The Institute for Rural America was trying to raise money to bring organic soybean producers from the Midwest together to discuss what they might do. The grant was not funded. However, by the time our funding fell through, I had learned that the problem of chronically low prices went beyond food grade soybeans. Low prices were the rule, not the exception. Even though we didn't get our grant, we tried an end run and sent out invitations to something like 100 organic and sustainable farmers and their organizations to meet in Ames, Iowa. The National Farmers Organization (NFO) was willing to host the meeting at its office.

We had no idea who, if anyone, would show up. As I recall, about 75 people came to Ames for the meeting. Since those attending knew me as the one who had sent the invitations, I facilitated the meeting. We got to know each other, learned more about our common problems, and agreed to meet again in a few months. About 50 people attended that second meeting and heard Marvin Beshore, an attorney from Harrisburg, Pennsylvania, make a presentation by speaker phone about farmers' rights under the Capper-Volstead Act. About half way through the meeting, I ceded my facilitator's job and became a note taker. Oren Holle, the farmer from Kansas I mentioned earlier, took over.

Our next meeting was at the NFO convention in Minneapolis, Minnesota. Some of the farmers from the original group met with two friends of mine who were involved in what was called a "marketing-agency-in-common" for Central Milk Producers Cooperative (CMPC) in Chicago. CMPC was an umbrella organization for all the dairy coops shipping conventionally-produced fluid milk to bottlers in the Chicago Federal Milk Marketing Order. CMPC had been formed to ensure that all the coops sold milk at the same price and that buyers made full and timely payments to the coops and their farmer/members.

From that meeting, some of the organic grain farmers thought we could use the CMPC model as a way of bringing together a group of organic grain coops. In subsequent conference calls, the handful of participants from those original meetings came up with a name: Organic Farmers' Agency for Relationship Marketing or OFARM. The emphasis would be on the word "relationship" in the name. Some of those relationships were between the organization and its member cooperatives and their organic farmer members. Other relationships would be developed with buyers and others in the marketing channels and, ultimately, with consumers.

I was asked to start working as OFARM's first executive director in 2001. We became officially incorporated in the State of Minnesota in 2005 as a Capper-Volstead qualified cooperative operating as a marketing-agency-in-common. OFARM has since grown into the largest single organized block of organic producers growing organic field crops in North America. We have six member cooperatives that serve hundreds of organic grain and livestock producers in 19 states.

A hallmark of OFARM is not only the marketing service functions it provides to its organic farmers, but detailed attention to sustainable organic market prices. After years of refinement, experienced organic grain farmers across the United States have worked through OFARM to develop five specific considerations for profitable organic grain production:

1. The full recovery of all actual production inputs, including those unique to the production, handling, and marketing of organic grain.

2. A return to labor and management that provides family income at levels that allow for the full involvement and adequate compensation of all members of this partnership in the operation of an organic production unit. This compensation must extend to the education, training, and transition to a future generation of organic farmers.

3. Return to investment that provides for the acquisition and ownership of the land and related infrastructure required for organic food production.

4. Income enhancement to provide for support for the social and economic viability of the community.

5. Organic price for production of healthy, wholesome food in an environmentally responsible manner.

These goals are incorporated into OFARM's list of target prices for grain and livestock that our farmer Board develops and updates on a regular basis. Those target prices, in turn, guide the marketers from the OFARM member cooperatives as they go into the organic market each day to negotiate and contract hay and grain on behalf of our farmer/members.

Oren Holle has often stated of OFARM, "We are more and more now writing the contracts rather than considering buyer's contracts. I'm talking about exclusively offering our organization's contracts to the various entities that might buy our products. It's difficult to say for sure how much effect we have had on the general market level. But there isn't any doubt in my mind that the system works for farmers. Take the wheat market, for example. The history of the wheat market is that every year we started over at the bottom. If it happened that there was a fairly short wheat crop, prices would creep up pretty good. But if we had a full crop of wheat it was never worth very much. Now we're moving price progress from one year to the next with contracting opportunities. If that's not a positive market impact, I don't know what is."

Dr. Thomas Gray, a USDA rural sociologist, wrote an article about farmers working together in the January/February 2012 issue of *Rural Cooperatives*. In that article, he said, "What large corporations cannot offer is member control and democratic governance. OFARM helps empower its member organizations and member farmers by facilitating scale, assembly, marketing and purchasing functions in order to improve viability of small local entrepreneurs."

Carmen Fernholz puts it this way, "It becomes one of the major missions of OFARM to constantly be in touch with producers in our member organizations. We show and demonstrate to them how the building of market strength is in fact the only option they have if they are going to maintain what they now have, but also for generations to come."

As for me, I can't say anything better than what OFARM now has for its mission statement, that is, "Establish and maintain sustainable prices for organic farm production through coordinated efforts of organic farmer cooperative marketing groups while protecting and defending the organic standards and promoting environmentally friendly production practices."

Our member cooperatives have hired the marketing agents you will need to be successful in today's organic grain markets. OFARM has linked those cooperatives together in ways that build strength in the market place. Now it's your turn. Will you join us and seek marketing expertise and strength in numbers, or will you take your chances on your own in a market dominated by corporate giants?

CHAPTER EIGHT

The Price Advantage
of Group Marketing

There's no getting around the "hassle factor" in marketing organic grains. Marketing organic grains takes more paperwork, more testing, more looking for buyers, more concern with shipment, and more worrying about payment than conventional grains. As I said in the previous chapter, most organic grain farmers will be better off dealing with these hassles through the services of an experienced marketing agent.

A good marketing agent can also help you find the best price available for what you have to sell. But here's something to think about: What makes the best price you can find, for example, $10 instead of $12? Often, the answer to that question depends on how much negotiating power you have. Buyers are almost always bigger than individual farmers, so they naturally have more negotiating power. But organic farmers joining together in group marketing can level the playing field and make the available prices better than they otherwise would be.

Group marketing is important for all farmers, not just organic farmers, or even just organic grain farmers.

Here's a good example from the *St. Louis Post-Dispatch*.[4] It's from a story about how McDonald's is planning to phase out certain antibiotics from the chicken the global giant buys from its suppliers. Production costs would be higher, and that raised the question of who would be stuck paying for those higher costs.

McDonald's didn't want to pass the costs along to its customers because the company was already facing stiff price competition from other fast food chains. McDonald's could therefore absorb the new costs as lower profits, or send the bill to its suppliers. Want to guess who came out on the short end of that one? The losers would be the suppliers who, according to the article, "may not have the market power to resist." McDonald's "carries a lot of clout with suppliers." In some cases, suppliers are completely dependent upon McDonald's to sell anything at all.

So, even though the suppliers are themselves corporate giants in many cases, they are no match for McDonald's when it comes time for head-to-head negotiations about prices and contract terms. I'll leave it to you to guess what chance an individual farmer has in such a bad economic neighborhood.

I'm sure many of you reading this are thinking, "But organic farmers don't sell to McDonald's, so what does this have to do with me?" Believe me, it has a lot to do with you. Giant food companies are taking over the retail

---

[4] Huffstutter, P.J. and Lisa B. Reuters. "Chicken growers set to pay price for no-antibiotic McNugget" *St. Louis Post Dispatch*. March 8, 2015.

markets for organic products of all types. This is happening so rapidly that whatever I write now will likely be old hat by the time you read it. In 2012, the *New York Times* put it this way: "Organic food has become a wildly lucrative business for Big Food and a premium-price means premium-profit section of the grocery store." Since then, things have gotten worse, not better.

There are always more farmers ready to sell their products than there are corporations to buy those products. This puts buyers in a strong position: they can keep prices relatively low by playing one farmer against another. The best solution is for farmers to level the playing field by acting together as if they were a single seller. That such actions are perfectly legal for farmers, and perfectly illegal for other businesses, is a great and underutilized advantage for farmers.[5]

You have all heard the time-worn advice to "get big or get out" at one time or another. Too many times, farmers have found that "get big AND get out" was more like it. Those farmers put all of their efforts into getting big, and none into getting the prices they needed to stay in business. When it comes to marketing, a better motto is "act big or get out." The Capper-Volstead Act makes it both legal and possible for independent farmers to "act big or get out" when they market their products. They can also do it without sacrificing their independence.

---

[5] Parts of the following text were adapted from: Levins, R. A. *Market Power for Farmers.* Institute for Rural America, Ames, IA. 2005.

The problem of corporate power being used against farmers is hardly a new one. In the late nineteenth century, a U.S. Senator urged action against powerful corporations that "increase beyond reason the cost of the necessaries of life and business and decrease the cost of the raw material, the farm products of the country." The Sherman Act, which became law in 1890, made it illegal for corporations to act together to set prices or otherwise manipulate markets.

There was a potential problem, however. Speaking about farmers, a Congressman told his colleagues in 1914: "One of the main objects is by cooperation to secure the best market and price for their products." Corporations could merge into larger, more powerful units, and thereby avoid the provisions of the Sherman Act. The farmer didn't have this option. Rather, as another member of Congress put it, the farmer "wants to be independent, but he wants to cooperate with other independent farmers in buying supplies and marketing his products without being under the ban of the law— without being criminal."

The Sherman Act appeared to make farmer cooperation illegal. This was clearly not the intent of Congress. The sentiments expressed by Representative Barkely on the House floor in 1920 ring as true today as they did then: "We all know that it is economically impossible for any individual farmer to compete with the conditions under which he must live. When he sells his product, he must sell it at a price dictated not by himself but by others who have had no part in its production." Two years later, the Capper-Volstead Act made the appropriate exceptions. Farmers could act together to raise capital for value added ventures, and they could act together in ways to effectively negotiate prices with powerful buyers.

For more than 90 years, the Capper-Volstead Act has waited patiently for farmers to apply it effectively in to-day's world of agribusiness giants. Laurance Waldoch, General Legal Counsel for the National Farmers Organization, sees the act as "more important than ever now that agribusiness has become so consolidated." His years of experience with Capper-Volstead issues have led him to believe that "groups of farmers strategically located near key buyers can be quite effective in building negotiating power."

Dr. Neil Harl of Iowa State University agrees that the law is very timely:

The 1922 legislation offers a framework for eligible persons to act together in achieving legitimate economic ends. In the current environment of towering concentration on much of the input side of agriculture and towering concentration in the handling, processing, shipping, storing or drying commodities, farm firms in nearly perfect competition are highly vulnerable to economic pressures from both sides. Capper Volstead . . . offers contemporary protection from antitrust charges if farmers organize to achieve a measure of counter-vailing power in the marketplace.

Unfortunately, *having* legal rights and *using* those rights effectively are two different things. The late Dr. Harold Briemeyer was my mentor and graduate professor at the University of Missouri. He once said something that really stayed with me:

> *Almost by their nature, family farmers lack powers of survival. The reason lies in the psychology of the individual farmer, whose concentration on his own operation tends to distract him from concern for forces that affect family farming as a whole. This has been referred to as family farmers' "non-instinct for self-preservation."*

Organic grain farming is relatively new on the American agricultural scene. Because of that, we have a choice to make. The lonely road of competing with our neighbors is one path we can take. That amounts to being organic farmers when we grow things, but conventional farmers when we sell what we produce. If we take that road, we can expect the same results that conventional farming has seen, that is, low prices and declining farm numbers. Or, we can use our Capper-Volstead rights by joining with our neighbors in building what Dr. Harl aptly calls "countervailing power in the marketplace."

At OFARM, we want more than a better way to produce food. We want a better way to market what farmers produce. We want to use our legal rights to bargain for prices that are fair for everyone. We want prices that allow farmers to stay in business and thereby support rural communities. We want prices that don't force farmers to cut corners on environmental protection. Isn't that what you want, too?

So, that sets the stage. Now let's get down to the specifics of working together as organic grain farmers to make sure we get our fair share of the market we have helped create.

Let's look first at market information. There is no organized, widely-available source of pricing information organic grain farmers can use to their advantage. Put another way, organic grain farmers don't yet have the CBOT and other resources available to conventional producers. Because buyers are usually bigger than organic farmers, they have better information. They see more producers and can play one against the other. And, let's face it, any time your principal source of market information is your buyer, you're in trouble. That applies to any business, not just organic grain farming.

One of OFARM's principal goals is that of collecting high quality market data that strengthens our bargaining position with buyers. The marketers from all of the member cooperatives regularly share information concerning product on hand, expected production, available pricing, and market opportunities. Collectively, the organized farmers are often larger than any one buyer and therefore have superior information. That information alone should translate into more profitable decisions for the group.

The advantage for organic farmers doesn't stop there, however. I'm sure you have heard the phrase, "Knowledge is power." That's true, but you have to use that power for it to be effective. OFARM puts the cooperative marketers in a better position to negotiate higher prices than any individual can get. Why? The marketers represent far more volume than any individual farmer can bring to market. The more volume you have, the more negotiating power you have. As simple as that sounds, it lets our participating organic farmers move from being price takers to price makers.

The strategy works. OFARM asked Dr. Richard Levins to do an independent study of our negotiating program during a two-year period. He compared prices received by organized OFARM farmers and by farmers marketing on their own.[6] For crop year 2002, OFARM farmers were paid higher prices for corn, clear hilum soybeans, Vinton soybeans, and spring wheat. The biggest payoff for group marketing in that year was for spring wheat; farmers in the OFARM-affiliated cooperatives got prices that were 42 percent higher than what farmers acting alone were able to get. For crop year 2003, OFARM prices were again higher for corn, clear hilum soybeans, Vinton soybeans, and spring wheat. OFARM price premiums of 24 percent for Vinton soybeans and 22 percent for spring wheat were reported.

Dr. Neil Harl of Iowa State University once said, "When competitive markets are compromised, classical

---

[6] Levins, R. A. "Midwest organic farmers see benefits from 'coop'-erating." *Leopold Letter*. Spring 2005. Iowa State University, Ames, Iowa.

market forces no longer determine prices. They are negotiated." I certainly think that the latest wave of mergers and acquisitions in the organic supply chain don't leave as much room for competition as we once had. If you agree, then you have to pay special attention to what Dr. Harl says. Hard work and smart management are no longer all you need to get the most from your organic grain farm. You also need negotiating strength.

So let me leave you with this question: "Are you better off negotiating with giant buyers on your own, or when you are working together with your fellow farmers to build market knowledge and volume?"

Here's how I would answer that question: JOIN AN OFARM COOPERATIVE!

The Other Side of the River

Back in 2002, just as I was getting started with OFARM, a University study came out that confirmed everything that had led me to this side of the river. Research by Dr. Luanne Lhor at the University of Georgia found that:

- Counties with organic farms have stronger farm economies and contribute more to local economies through total sales, net revenue, farm value, taxes paid, payroll, and purchases of fertilizer, seed, and repair, and maintenance services.

- Counties with organic farms have more committed farmers and give more support to rural development with higher percentages of resident full-time farmers, greater direct-to-consumer sales, more workers hired, and higher worker pay.

- Counties with organic farms provide more bird and wildlife habitat and have lower insecticide and nematicide use.

- Watersheds with organic farms have reduced agricultural impact and lower runoff risk from nitrogen and sediment.

- Organic farming methods and prices support the production of socially desirable outputs, such as clean water, as a byproduct of food production activities. This reduces the need for government intervention through taxes or subsidies to obtain these benefits.

Then, like now, I knew that organic farming was better for farmers, their families, and their communities. But I also knew that organic prices would not magically stay at levels that would lead to growth. I worried that organic farmers would fall into the same economic traps that had ensnared so many of their conventionally-farming neighbors. How to fashion an organic market that would support sustainable growth in organic farming had been on my mind for years. At OFARM, I was able to put those ideas into practice.

I suppose every year has its ups and downs. The year 2002 was certainly like that for me. I was happy to see the release of Dr. Lohr's study and to watch OFARM gain its footing. But I was deeply saddened by the death of my graduate school professor and long-time friend Dr. Harold Briemyer. His thoughts, as much as any, have guided me in my search for ways to build fair markets for farmers. For the record, I am deeply respectful of the merits of the family farm as a business unit and of the family farm itself. But I also agree with Dr. Briemyer:

the family farm is in jeopardy today precisely because its principal beneficiaries—farmers and their families—have been preoccupied with their self-concerns. They, like almost all of their non-farm supporters, have failed to understand what is necessary to preserve, not the individual farm, but family farming as a system.

Over the last 40 years of my career in agriculture, I've seen conventional farming trade its soul for a mantra of "be the most efficient." In my darker moments, I don't see a way for conventional agriculture, now almost completely captured by giant agribusiness corporations and industrial thinking, to recover. But organic farming gives us a second chance to get it right. We have the remarkable good fortune to learn from past mistakes and fashion a farming system that preserves family farming, our environment, our communities, and our health.

My vision of rural America is one that has a place for everyone at the table—farmers and their families, the industries that support them, and consumers who share a vison of what we want our communities, farms, and families to be in the future. I'm optimistic that growth in organic agriculture will help us achieve that vision. Sure, organic farming, just like anything else worth doing, will always be a struggle. But younger farmers, if they heed the lessons and failures of conventional agriculture, have the opportunity to forge a bright future not only for themselves, but for all of us who depend on farmers every time we sit down to eat a meal.

Crossing the river from conventional farming to organic farming has taken me a lifetime. I hope our younger farmers will make the journey more quickly. But no matter how long the journey takes, I know that each and every one of those who make it will feel as I do. The journey, no matter how difficult, is more than worth it. The view from the other side is truly wonderful.